Learning Logic

Critical Thinking with Intuitive Notation

Reviews

"The theory of logic, the meaning of logic, the study of logic, and the practice of logic are all dependent upon the language of logic. Without a uniform language, logical discourse is fraught. This book is not only a clear, straightforward text on the building blocks of logic, but also a handbook which proposes a standardised means of describing those building blocks. As such it is the sort of publication that will be used frequently and well thumbed. Plowright has taken a commonsense approach to developing a clear, intuitive and well explained uniform terminology for the language of logic. The significance of this book is that it enables logical argument with common understanding, thus facilitating sound, well-reasoned logical conversation. That is a major contribution to the field for a humble little book."
 -- *Associate Professor Brad Mitchell,*
 Principal Research Fellow,
 Federation University Australia.

"Plowright has succeeded admirably in achieving his aims of writing this book. He has whetted the appetite of a thinking person to further explore the tools of formal logic. The range of situations he has selected illustrates the need for critical thinking in almost every area of human endeavour. His introduction to techniques such as Venn diagrams and truth tables is clear enough for the reader to follow, and with a little work, be convinced of their validity in solving logical conundrums. His brief introduction to basic probability theory convinces the reader that further study in this area will be invaluable in refuting claims often based on misleading evidence, particularly where statistics is involved.
 In all, a worthwhile and informative read which opens the door to an area of mathematics of great value but as yet, undiscovered by many."
 -- *Michael Fenelon*
 Senior High School Mathematics Teacher, Victoria Australia

"In a world where politicians and celebrities have elevated fuzzy thinking to an art form, this book will show you how to cut through the fog, sharpen your critical thinking skills, and avoid the logical pitfalls that lie in wait for us every day."
 -- *R. S. Radford,*
 The Radford Center for Law, History & Economics, California USA

This Book is Available From
www.lulu.com & www.amazon.com
and other online bookstores

Learning Logic

Critical Thinking with Intuitive Notation

Stephen Plowright

London – Toronto – Internet

First Published in 2015
by Lulu Press

www.lulu.com

Copyright © Stephen Plowright 2015
Printed digitally

The moral rights of the author have been asserted.

All rights reserved. No part of this publication, either in part or in whole may be reproduced transmitted or utilised in any form or by any means electronic, photographic or mechanical, including photocopying, recording, or by any information storage and retrieval system, without the permission in writing from the Author, except for brief quotations embodied in literary articles and reviews.
All enquiries should be addressed to S. Plowright, steve9@mackaos.com.au

This book may not be circulated in any other binding or cover.

ISBN 978-1-329-44306-8

Contents

Introduction — 1
Logic, what is it good for? — 3
Faulty Arguments — 10
Propositional Logic — 15

Propositions — 15
Operators — 17
Conjunction, AND, ∧ — 18
Disjunction, OR, ∨ — 20
Negation, NOT, ¬ — 22
Equivalence, EQUIVALES, ≡ — 24
Implication, IMPLIES, → — 26
Entailment — 29
Combined Operators — 30
More about Logical Equivalence — 32
Boolean Variables and Mathematics — 34
Axioms and Theorems — 35
Truth Tables — 37
De Morgan's Law — 39
Proofs — 40

Predicate Logic — 41

Quantification	42
Type & Domain	44
Universal Quantifier, ∀	45
Existential Quantifier, ∃	46
Notation	47
Bound variable	52

The Logic of Sets — 53

Venn Diagrams	55
Set Operators	56
Set Membership, ∈	57
Union, ∪	58
Intersection, ∩	59
Complement, ~	60
Subset, ⊆	61
Proper Subset, ⊂	62
Set Equality, =	63
Set Difference, -	64
Special Sets	65

Logic and Uncertainty — 66

Testing and Detection — 66
Probability — 69
Simple and Compound Events — 72
Sets in Probability — 74
Conditional Probabilities — 77
Bayes' Theorem — 81

Putting Logic to Use — 82

Propositional Puzzle — 84
Predicate Puzzle — 86
Quantification Puzzle — 88
Logic and Probability Puzzle — 89
Events and Detection Puzzle — 92
Classic non-Paradox of Probability — 96
Rolling Dice — 99
Common Sense and Science — 102
Further Reading — 110

Acknowledgements

I would like to thank and acknowledge the unfailing support and advice of my wife Jodie. I am also grateful to my colleagues and managers at IBM, who were the guinea-pigs who test-drove the short course upon which this book was based, and to the Ulysses motor cycle club for their support and proof reading.

In every country, we should be teaching our children the scientific method and the reasons for a Bill of Rights. With it comes a certain decency, humility and community spirit. In the demon-haunted world that we inhabit by virtue of being human, this may be all that stands between us and the enveloping darkness.

Carl Sagan

Introduction

Logic is the study of the mechanics of reasoning. It is a method of understanding how we make sense of the world. To some extent, the ability to think logically is innate, or at least seems to be. Thus, our ability to understand arguments, or puzzles, is often taken for granted.

However, the more complex, or unfamiliar the reasoning becomes, the more mistakes we make by relying on our intuition or intelligence alone.

Formal logic provides an x-ray view of the skeleton of a line of reasoning. It does this by stripping away the content and representing the relationships between the elements of the argument with symbols.

These symbols and their usages are what we call the notation. The intention of this book is to present an intuitive and consistent notation that follows from one area of logic to another, while being consistent with the familiar notation of mathematics.

By optimising the flow of the notation, the underlying principles of logic are more easily assimilated. The familiar aspects and simple mnemonics make learning and remembering the symbols easy, which in turn makes the underlying logic easier to perceive.

To assist further with the understanding of principles, logical components will be illustrated, where possible, in several contexts: linguistic, symbolic, mathematical, and physical. Each chapter lays the foundation for the next.

It is my hope that this manual will provide a basic introduction for anyone interested in any of the applications of formal logic, from argumentation to programming, from critical thinking to science and mathematics.

It is not meant to be a thorough textbook on logic, and it will require some effort to make the connections, but each section will be understandable in terms of previous sections. If approached as a sequence of building blocks, it will familiarise the reader with concepts that will make further study easier, while giving the casual reader some useful insights into critical thinking.

Chapter 1
Logic, what is it good for?

The basic operations of logic are performed by all of us in an almost unconscious way. This has led many philosophers to assume that it is something innate, and that the rules of logic are universal truths that are self-evident. Others argue that we learn the rules of logic as children, through interacting with the world.

Despite the feeling that most of us have, that we can put forward or comprehend a logical case, most people are prone to making mistakes when the example is complex or unfamiliar.

About 23 centuries ago, Aristotle was observing the arguments of philosophers in ancient Greece. He realised that there are some arguments that make sense. They start with some truths, and always come to a true conclusion. Also, that there are some arguments that do not work. They may start with truth, but often lead to absurd conclusions. He realised that in order to understand why, he had to separate the form from the content. The problem was not what people were arguing about, but the way they were arguing.

He set out some forms of reliable arguments he called "syllogisms", and some faulty forms called "fallacies".

The best-known classic form of syllogism is: All humans are mortal, Socrates was human, therefore Socrates was mortal. If the first two statements (the premises) are true, then the third (the conclusion) **must** be true.

An argument that always has a true conclusion, as long as its premises are true, is called a **valid** argument. A valid argument with true premises is called a **sound** argument. Aristotle remained a primary text for logic for the next two thousand years.

Meanwhile, another Greek thinker, the mathematician Euclid, collected the many known geometrical proofs, including the famous theorem of Pythagoras. He added some proofs of his own, and it could have ended up little more than a collection of mathematical exercises. However, he went much further, and introduced a framework at the beginning of the book, defining the most basic elements and rules of geometry, the **axioms**. This work remained the primary textbook in geometry for the next two thousand years.

And it was two thousand years later that the German polymath, Leibniz, realised that the forms of arguments described by Aristotle, could be broken down into basic parts and rules, like the axioms of Euclid's geometry. Thus the foundations of modern axiomatic logic were laid.

However, Leibniz never published these ideas. It was almost two centuries later that they became fully realised in the "Boolean Algebra", developed by Boole and De Morgan in the mid 1800s.

This is only a very basic outline of the historical background of formal logic. Boole used symbols like + and =, which already existed in mathematics, to represent logical operations, and this could cause some confusion for students. We will use unique symbols, but with the same consistent and symmetrical qualities that we see in mathematics.

In renaissance times, there were few conventions in mathematical notation. Each textbook or paper had to define the symbols it would use. Today, we take it for granted that everyone uses the same symbols for addition, multiplication, and equality.

Logic, however, has not yet reached this level of consistency, and you will still see a different mix of symbols defined in each book. I have chosen the most consistent and symmetrical out of the commonly used symbols to use in this book. This is mainly the notation used by Gries & Schneider in their 1993 computer science text "**A Logical Approach to Discrete Math**", and Prof. Stan Warford in his iTunesU course on "**Formal Methods**".

Learning logic can benefit anyone. It is good exercise for the mind, but also teaches us to be able to understand the reasoning behind arguments in philosophy and law, or the logic behind the methods of science. It allows us to see through the illogical claims seen commonly in politics, fundamentalism, and advertising. Perhaps most importantly, it makes us better able to make a credible case when putting forward our own ideas, both in higher education, and at work.

Logic is essential for making sense of the world. The reason we can survive is that we are able to become familiar with the consistent behaviour of the world. We can predict the consequences of actions, or the progress of a chain of events. We are, in effect, our own scientists.

Science is the most successful means yet found of learning about the world. The logical and predictable nature of the world is the basis of the scientific method, which utilises two forms of reasoning: inductive and deductive.

Deductive logic is the type we have described already, where true premises lead to true conclusions.

About 400 years ago, Sir Francis Bacon, an English lawyer, realised that deductive logic was important in drawing conclusions from established truths, but was not equipped to find new truths. His work on **"The Advancement of Learning"**, laid the foundations of the modern scientific method by introducing inductive logic.

Induction was described as the method of looking for the consistent patterns or laws in nature. By observing and experimenting, we can collect data to reveal these laws, and allow us to use the laws to make predictions about the consequences of new situations. The result was an explosion of scientific discovery in the following centuries, and the "Age of Reason".

The very success of science in discovering and utilising the laws of nature, and its spinoff technologies, are ubiquitous testaments to the power of inductive and deductive logic in making sense of the world.

In everyday life, logic is vital in evaluating the credibility of information, drawing conclusions from the available information, and making good decisions and plans.

We are surrounded by illogical claims. Pseudoscience, fundamentalism, advertising, and politics, inject emotion-laden messages to appeal to our fears and desires.

Parents are making health decisions for their children based on anecdote and rumour, amplified and re-broadcast by social media. The

Internet gives us access to an unprecedented amount of good information, mixed with an even greater quantity of complete rubbish.

Pseudoscience, charlatanry, conspiracy theories, and denial masquerading as scepticism, are not victimless crimes. They have direct harmful consequences on the outcomes of decisions made or delayed, based on misinformation.

Poor reasoning skills prevent people from recognising the signs that indicate the suspect assumptions and absurd conclusions that litter the Internet. Even when accurate raw data is obtained, many educated people draw erroneous conclusions, especially when probabilities or numbers are involved.

A basic understanding of logic is invaluable in achieving our goals in life, and ambitions at work. In law, science, technical, and engineering careers, logic is explicitly required, although often not formally taught.

With a fairly modest effort, a substantial reward of clarity can be achieved, and many mistakes avoided.

Chapter 2
Faulty Arguments

Fallacies, or arguments that are not logical, are particularly common in politics, fundamentalism, pseudoscience, and advertising. There are many forms, some more obvious than others, but once familiar with them, they are much less likely to catch us out.

First, let's start with the classics. These are not so much logical errors, as complete avoidance of logic.

Ad Hominem: Attacking the person instead of the argument. Very common in Internet "flame wars".

Tu Quoque: Answering a criticism by criticising the opponent.

Appeal to Authority: Claiming that a respected person backs your position.

Shifting the Burden of Proof: Making a claim, and insisting that it is up to someone else to disprove.

FAULTY ARGUMENTS

The Straw Man: Taking the opponent's argument and twisting it, then attacking the misrepresented version of it.

Appeal to Nature: Claiming that something is "natural", therefore good, or "unnatural", therefore bad.

Anecdote: Using specific isolated examples to dispute generally established or statistical facts.

Compromise: Agreeing arbitrarily that the truth lies between two positions being argued.

Obfuscation and ambiguity: Using words so that the meaning can be taken in more than one way, or to sound convincing without really meaning anything.

Cherry Picking: Choosing data and examples to support one case, while ignoring any evidence to the contrary.

Black/White: Only accepting two possible positions as valid. "If you are not with us, you are against us".

Appeal to Purity: Claiming that each counter example is not a pure or true case. Also known as the "no true Scotsman" argument.

Appeal to Emotion: Using emotional arguments to induce pity, guilt, shame, fear, or anger, to influence opinions.

Loaded Question: Asking a question that assumes fault in the opponent. "Have you stopped taking drugs?"

Fallacy on Fallacy: Asserting that the conclusion of an argument must be wrong just because the argument was faulty.

Appeal to Bandwagon: Claiming that a statement is true because "everyone" says so.

False Causality: Assuming that a correlation between two events means that one caused the other. "Most grandparents have grey hair, therefore having grandchildren causes aging".

So far, we have seen argument styles that do not even get as far as using logic. Now we need to think about arguments that use faulty logic. These can be harder to spot.

Inclusive/Exclusive OR: Making a reasonable claim of "A or B" being true, then showing that A is true, and basing the conclusion on an assumption that B is false. This argument is only valid if A and B are mutually exclusive, (ie can't both be true). In the normal usage of OR, both A and B may be true.

For example, "Jim will not ride his bike only when (A) it is raining or (B) it dark. Dave does not care about (A) the rain, but will not ride when (B) it is dark". Jim sends a message that he will not ride. Dave now knows that either it is raining or it is dark. He looks on the weather radar and confirms that it is raining. Can he assume that it is ok to ride (not dark)? However, if he can confirm it is not raining, then he can assume it is dark.

Reversed Implication: Making the reasonable argument that if A is true then B must be true, but then claiming that since A is false, B must be false. Also fallacious is the claim that if B is true, A must be true.

For example, "If there are germs in the water, it is unsafe to drink. We tested and found no germs, therefore it is safe". This is the most common kind of fallacy you hear in politics and advertising. The germs may have been killed by high levels of cyanide. The implication only goes one way.

Incorrect Generalisation: Making a reasonable assumption about a specific collection of things, and then basing the conclusion on a subset or superset of the original collection. This fallacy is often combined with a Straw Man attack in attempts at political correctness. For example, I say, "My Dutch relatives are outspoken", which receives a self-righteous response "You can't say that about the Dutch".

Chapter 3
Propositional Logic

Propositions

A proposition is a statement that is either true or false, but not both at the same time. "The sky is blue" may be true on a nice day. "The moon is round" may always be true.

Like numbers in mathematics, propositions in logic can be represented by a lower case letter in order to strip away the content and reveal the relationships between the propositions.

Rather than representing a large number of possible numerical values, as in algebra, propositions are variables with only two possible values.

If the variable **p** has only two possible states, it is called a Boolean variable. The two values can be (0, 1), (off, on), (black, white), or any mutually exclusive pair of conditions, depending on the application. For our purposes, (true, false) is the most useful pair of values.

If we let **p** represent the proposition that "the grass is green", then we can assert or deduce whether **p** is true or not true, according to the logic of the argument, without regard to any preconceived notions we might have about grass.

Operators

Mathematical operators are used to perform operations on or between numbers. We are familiar with addition, multiplication, etc. We are also familiar with the symbols +, ×, =, etc.

Logical operators are used to perform operations on or between propositions. They are expressed in natural language by words such as "and", "or", "not", "implies", etc.

We will now explore the symbols for logical operators and their usages. This is the formal notation that will enable us to reveal the logical structure of arguments.

Conjunction, AND, ∧

If we have two propositions, we can combine them with the word "and".

Let **p** stand for "it is raining". Let **q** stand for "the road is wet". We can represent the combined statement as "it is raining **and** the road is wet". The combined statement is understood to be true only if both parts are true.

We use the symbol ∧, which is easy to remember as resembling A for "AND".

We use this in the following way:

p ∧ q

to signify the statement "**p** and **q**", where **p** and **q** are propositions, and (p∧q) is true only when **p** and **q** are both true.

We can also construct a truth table for all possible values of **p** and **q**:

p	q	p ∧ q
True	True	True
True	False	False
False	True	False
False	False	False

Physically, we can represent the behaviour of the table with a circuit with two switches in series.

Both switches are off (breaking the circuit).

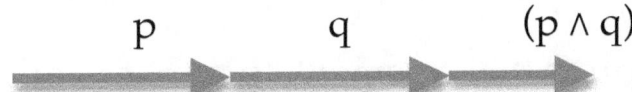

Only when both switches are on, can current flow.

If either switch is off, the path is broken, and current cannot flow.

Disjunction, OR, ∨

Two propositions can be joined with "or". The effect is different from "and". To use the propositions from the previous example, we can imagine not wanting to go for a ride when "it is raining, **or** the road is wet". Now the combined statement means when either **p** is true, or **q** is true, or when both are true.

The symbol is ∨, and memorable as being like the "AND" symbol but upside down.

We use the symbol as follows:

p ∨ q

to express the statement "**p** or **q**, or both".

Note: It is not to be confused with the "exclusive or", which means "either one or the other, but not both".

We can also construct a truth table:

p	q	p ∨ q
True	True	True
True	False	True
False	True	True
False	False	False

Physically, we can represent the behaviour of the table with a circuit with two switches in parallel.

The circuit is broken with both switches off.

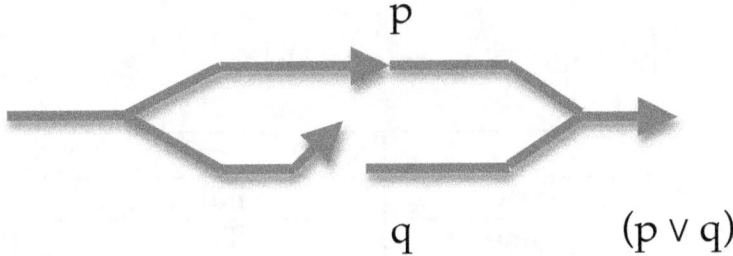

The current flows if either one, or both switches are on.

Negation, NOT, ¬

If we use the example of **p** being the proposition that "it is raining", we will also want to be able say "it is not raining". Only one of these statements can be true, as each is the negation of the other.

"NOT" will reverse the truth-value of the proposition. It reminds us of the negative sign in mathematics.

It is used as follows:

¬p

to represent "not **p**", the negation of **p**.

It has the following truth table:

p	¬p
True	False
False	True

We can see that if we negate **p** twice (¬¬p), it will not change. In other words (not (not **p**)) is **p**.

A physical example is a switch with a linkage to another switch, such that they are always forced to be in opposite states. There is always one on & one off.

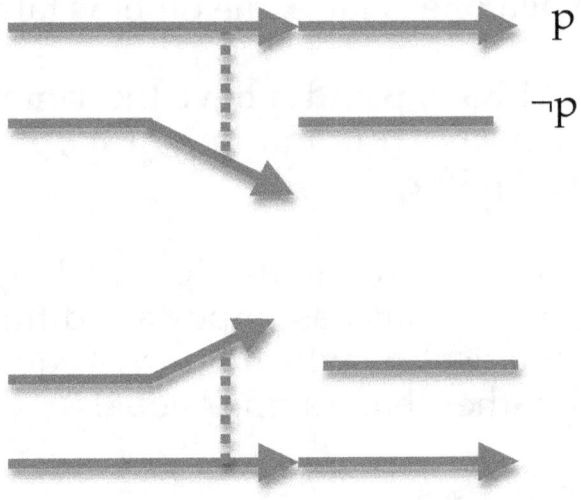

Equivalence, EQUIVALES, ≡

Logical equivalence exists between two propositions if when one is true, the other is true, and when one is false, the other is false.

If **p** EQUIVALES **q**, **p** and **q** have the same truth-value. We can say "**p** if and only if **q**". It is also often written "**p** iff **q**".

It looks similar to an equals sign, and signifies equal truth-value, but has important differences from the numerical = , which is why "equivales" was coined, rather than saying "equals".

It is used as follows:

p≡q

It has the following truth table:

p	q	p≡q
True	True	True
True	False	False
False	True	False
False	False	True

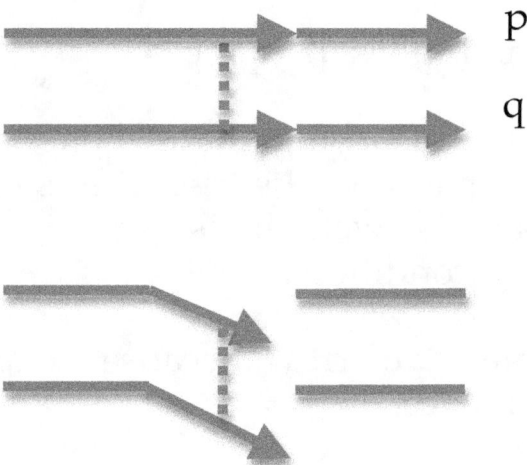

A physical example is a switch with a linkage to another switch, such that they are always forced to be in the same state. They are always both on or both off.

Implication, IMPLIES, ⇒

(If p then q), (p implies q), (q if p).

These are ways of saying that whenever **p** is true, we know that **q** will be true. **If** it is raining, **then** the road is wet. It is known as an implication or a conditional.

It is memorable as an arrow, because it expresses a **one-way** influence

It is used as follows:

p ⇒ q

Means "if **p** then **q**", or "**p** IMPLIES **q**".

(**p** is called the **antecedent**, **q** is the **consequent**)

If **p** is true, then **q** is true. Truth table:

p	q	p ⇒ q
True	True	True
True	False	False
False	True	True
False	False	True

A physical example would be a switch with a linkage that is only connected at one end. It can push the other switch to the on position, but not pull it back to off.

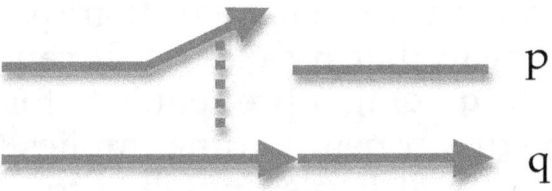

While the top switch is off, the bottom switch is free to be on or off. While the top switch is on, the linkage pushes the bottom switch down into the on position.

It is worth spending some more time on implication, as it is usually at the centre of arguments, and often misused.

As mentioned in the faulty arguments chapter, the one-way nature of conditionals means we must keep track of the direction of the implication, hence the arrow.

If we accept the statement $(p \Rightarrow q)$, then **p** being true guarantees the truth of **q**. It says nothing about the implication of not-**p**, so we can deduce nothing from the case where **p** is false. In other words **p** being true guarantees that **q** is true, but **p** being false can mean anything (**q** is true or false).

It is also important to remember that (**p** implies **q**) does **not** mean that **p** causes **q**. It can just as easily apply to **q** being a pre-requisite for **p**. For example "the driver being drunk implies that he has had alcohol". It is also possible to have an implication with no causal connection at all. For example "if the moon is full, we eat outside".

(If p then q) is merely a statement that there is a conditional relationship. We say **p** is a **sufficient condition** for **q**. In other words, all it takes is for p to be true to know that q must be true.

We can also say that **q** is a **necessary condition** for **p**. In other words, we know that **q must** be true for **p** to be true.

If **p** implies **q**, we can also deduce that if **q** is false then **p** cannot be true. This is called the **contrapositive**. $(p \Rightarrow q) \equiv (\neg q \Rightarrow \neg p)$

Entailment

One of the central methods in logical argument is known as **Modus Ponens**. It says if we know that **p** implies **q**, then knowing **p** guarantees the truth of **q** (p **entails** q).

$$p \Rightarrow q$$
$$p$$
Therefore q

Note that (**p** entails **q**) goes a significant step further than the statement (**p** implies **q**).

We can also use the contrapositive to prove a negation.

$$p \Rightarrow q$$
$$\neg q$$
Therefore ¬p

This is called **Modus Tollens**.

Combined Operators

Sometimes it can be useful to use shorthand for combined operations. For example, non-equivalence, exclusive OR, and implied by.

Non-equivalence $\not\equiv$:

$\neg (p \equiv q) \equiv (p \not\equiv q)$

not-(**p** equivales **q**) is (**p** not-equivales **q**).

Exclusive OR can be written as $\underline{v}$, or as $\oplus$, or XOR. Where we want **p** or **q** , but not both:

$((p \vee q) \wedge \neg (p \wedge q)) \equiv (p \underline{v} q)$

(**p** or **q**) and not (both **p** and **q**), is (**p** XOR **q**).

However, if we look at the truth-values, we can see that non-equivalence requires **p** and **q** to always have the opposite truth-value. XOR says one or the other is true, but not both. Therefore, the XOR and non-equivalence have the same truth table, and are in fact the same operation. This comes as a surprise to most.

$(p \not\equiv q) \equiv (p \veebar q)$

(**p** not-equivales **q**), is the same as (**p** XOR **q**).

We will use $\not\equiv$ for non-equivalence, as the most intuitive symbol, being similar in look and meaning to the familiar $\neq$ in mathematics.

For implication, $\Rightarrow$, we have seen "**p** implies **q**", or "if **p** then **q**". We also have the option of saying "**q** is implied by **p**", or "**q**, if **p**":

$(p \Rightarrow q) \equiv (q \Leftarrow p)$

Which may be helpful if translating text as directly as possible to logic.

There is also a logically equivalent way to express implication that can be handy in proofs:

$(p \Rightarrow q) \equiv (\neg p \vee q)$

Try constructing a truth table for ($\neg p \vee q$), and compare it to the one we already have for ($p \Rightarrow q$) on page 26.

More about Logical Equivalence

When two propositions are logically equivalent, they always have the same truth-value. If **p** implies **q**, and **q** implies **p**, then they will always have the same truth-value. Hence the terms "bi-conditional", "if and only if", or "iff", are used for equivalence:

$((p \Rightarrow q) \land (q \Rightarrow p)) \equiv (p \equiv q)$

Some texts use (p ⇔ q) to show equivalence, emphasising the bi-conditional.

Another definition of ≡ :

$((\neg p \land \neg q) \lor (p \land q)) \equiv (p \equiv q)$

((not **p** and not **q**) or (**p** and **q**)) is (**p** equivales **q**)

There are good reasons to use ≡ rather than =.

They are very similar in most respects, but logical equivalence behaves a little differently, because mathematically, propositions are Boolean variables, as they can have only 2 values, unlike numerical variables.

For instance:

$p \equiv q \equiv q \equiv p$

This statement is always true no matter what the truth-values of **p** and **q** are. If **p** and **q** have different values then ($p \equiv q$) is false, also ($q \equiv p$) is false, but then (false $\equiv$ false) is true.

The expression ($p \equiv q \equiv q \equiv p$) is a **tautology** or **theorem**, an expression that is true in all states (all combinations of true and false for **p** and **q**).

Contrast this with = and numbers:

a=b=b=a

This can only be true when a=b.

Note also, that if $\mathbf{p} \equiv \mathbf{q}$, it does not mean **p** has the same meaning as **q**, only that it has the same truth-value. Even though they are different propositions, one can replace the other in a logical argument without changing the truth-value of the conclusion.

Boolean Variables and Mathematics

When we think of an expression like y=2x, we tend to think of it as "let y be equal to 2 times x", which then describes a line on a graph.

However, we can also consider that = is an operator that returns a Boolean result. The = is true when what is on each side matches, and false otherwise. The expression y=2x is a proposition that is either true or false, a Boolean.

The line defined by the expression is the region where y=2x is true, and the rest of the graph is where y=2x is false.

In programming particularly, we have to be very explicit about the difference between "let a=b", and "does a=b ?".

The mathematical operators <, >, ≤, ≥, ≠, and =, are all operators that return a Boolean result, true or false.

Axioms and Theorems

The **axioms** of a system are the most basic definitions and rules, and can only be accepted on trust. They may make sense, and work in useful ways, but by definition, they cannot be proven.

More complex statements can be analyzed using the axioms, and be proven true. If a statement is proven to be true in all states (for any combination of truth-values for **p**, **q**, etc), then it is a **tautology** or **theorem**. Once proven, theorems can then be used in the same way as axioms in further proofs.

Examples of axioms include **Excluded middle:**

¬ true ≡ false, ¬ false ≡ true

p ∨ ¬p

(**p** or not-**p**) is a statement that must always be true if we accept that there can only be two truth-values, (true & false). No matter whether **p** is true or false, either **p** or not-**p** has to be true.

Excluding a middle value makes **p** a Boolean variable (only 2 possible values).

Other axioms include:

Axioms of associativity: $(p \wedge q) \wedge r \equiv p \wedge (q \wedge r)$

Axioms of symmetry: $(p \wedge q) \equiv (q \wedge p)$

These are just a few examples of axioms. Most seem like common sense. I will leave it for the textbooks to provide an exhaustive list.

Axioms and theorems can be used to prove or construct further theorems.

Truth Tables

Now that we are familiar with the truth tables for simple operations, we can extend them to work for more complex statements. We can use them to verify the logic of arguments.

For example, to prove by truth table that non-equivalence and XOR are the same:

$(p \not\equiv q) \equiv (p \veebar q)$

We know $(p \veebar q) \equiv ((p \vee q) \wedge \neg (p \wedge q))$, which is just the definition of exclusive OR, (**p** or **q** but not both).

Constructing the truth table, with the results aligned beneath the operator that produced them, gives:

p	q	p ∨ q	p ∧ q	¬ (p ∧ q)	(p ∨ q) ∧ ¬ (p ∧ q)
t	t	t	t	f	t∧f ≡ **f**
t	f	t	f	t	t∧t ≡ **t**
f	t	t	f	t	t∧t ≡ **t**
f	f	f	f	t	f∧t ≡ **f**

The XOR expression is false if **p** and **q** have the same truth-value, and true when they are different, which is exactly the definition of (p≢q).

We can see that for two propositions, we get four possible combinations of values. For each further proposition, the number of combinations will double. If we had a statement involving **p**, **q**, & **r**, we would need 8 rows in the table (**n** propositions, means 2^n rows).

Truth tables are not the best method of proof for statements of more than three propositions. For such cases, we would use the method of using laws or rules, such as De Morgan's, to transform expressions on one side of the equation until they are the same as the other side, as we would in mathematics. We will now look at De Morgan's Law, and then prove the same example as above, using this method.

De Morgan's Law

$\neg(p \lor q) \equiv (\neg p \land \neg q)$

$\neg(p \land q) \equiv (\neg p \lor \neg q)$

Not (**p** or **q**) is the same as (not **p** and not **q**).

Not (**p** and **q**) is the same as (not **p** or not **q**).

De Morgan's Law is often handy in proofs, and also for simplifying search filters, where you are looking for entries containing some features but not others. For example searching for items containing **p** but not (**q** or **r**), can be transformed into **p** and not **q** and not **r**.

Proofs

In order to verify that an argument is valid, a proof can be performed. As in mathematics, it usually consists of reorganizing one side of an equation within rules that preserve the truth of the expression, until it is the same as the expression on the other side of the equation. I will leave the details of proof methods for the textbooks, but here is the example proved earlier by truth table. Prove: $(p \not\equiv q) \equiv (p \underline{\vee} q)$

$((p \vee q) \wedge \neg (p \wedge q))$:Definition of $\underline{\vee}$, XOR

$\quad = \quad \neg \neg ((p \vee q) \wedge \neg (p \wedge q))$:Double negation

$\quad = \quad \neg (\neg (p \vee q) \vee \neg \neg (p \wedge q))$:De Morgan

$\quad = \quad \neg ((\neg p \wedge \neg q) \vee \neg \neg (p \wedge q))$:De Morgan

$\quad = \quad \neg ((\neg p \wedge \neg q) \vee (p \wedge q))$:Double negation

$\quad = \quad \neg (p \equiv q)$:Definition of $\equiv$

$\quad = \quad (p \not\equiv q)$:Definition of $\not\equiv$

QED

The "=" here signifies a transformation that does not change the truth of the expression.

Chapter 4
Predicate Logic

So far, we have been able to deal with single propositions. Many arguments, both ancient and modern, deal with generalisations, or statements about groups of things. Much of Aristotle's logic involved categories, hence the phrase "categorically true" to express certainty.

Using propositional logic, if we have a room of people, who all belong to a club, we would have to say "Person A belongs to the club, and B belongs to it, and C...". Each person has a separate proposition:

a ∧ b ∧ c ∧ ...

We would like a more efficient way to say, for instance, that all of the people in the room belong to the club.

Quantification

Certain types of operation can be done in a "batch" fashion. You can add a bunch of numbers, and it makes no difference in which order you do it. This is like bringing your shopping to the checkout unsorted, knowing the total will be the same in any order. The same goes for multiplication. It turns out that "AND" & "OR" can also be batched. We can think of statements involving several propositions, and know that the order doesn't matter. For the mathematically inclined, this is an Abelian property of operators which include $+, \times, \wedge, \vee$.

You may be familiar with the symbol $\sum$ which signifies in mathematics that some numbers will be added together. Similarly $\prod$ is used when multiplying a list of numbers.

In order to apply such a process to propositions, we first have to split them into subject and predicate. In our example, the subject is the person, and the predicate is "belongs to the club". In general, a proposition will be of the form: **s** is **P**. Where **s** is the subject and **P** is the predicate. We can represent this as **P(s)**.

Now we can work with many subjects with respect to a single predicate:

$s_1, s_2, s_3, s_4, s_5 \ldots, s_n$ are all **P**
=
$P(s_1) \wedge P(s_2) \wedge P(s_3) \wedge \ldots \wedge P(s_n)$

Examples:

Bill, and Jane, and Ted, …., and Jill, are students.

2, and 4, and 6, ….etc, are even numbers.

Predicates form the basis of categorisation. A predicate or attribute **P** can create a category of subjects **s**. This is the category of the subjects that are **P**. For example, if **s** are people, and **P** is "working for IBM", we can use **P(s)** to describe the category of "people who work for IBM".

We will use these concepts in our discussion of quantifiers, but first we need to briefly look at how to define what subjects are relevant to an argument.

Type & Domain

An important concept in logic is the boundary of relevance around the things being discussed. This is commonly called the **Domain of Discourse**.

If we are discussing the people belonging to a company, we may not want someone to start bringing the furniture into the conversation, just because it also belongs to the same company.

To avoid these problems, we make the domain clear at the beginning. We can define the scope or domain of the argument by identifying the relevant subjects by **type**.

In our example, we can say the type is "people". If we are discussing numbers, the type may be "integers only".

Types can be represented with double-strike capitals: Booleans – $\mathbb{B}$. Integers – $\mathbb{Z}$.

Universal Quantifier, ∀

If we want to express the idea that all of the people are in the club, we are saying that one person is, AND the next is, AND …etc.

We are ANDing propositions with different subjects, but the same predicate. If we represent the subjects (people) with a variable "**x**", then we can say all **x** are **P**. We can use the universal quantifier symbol to represent that statement:

∀x P(x) Meaning: for all x, x is P.

For all the people, they are in the club.

We can memorise the symbol ∀ as an upside-down A for "All" or "Any", also "And".

We could also use the AND symbol to express the same idea, some texts do. It reminds us that "All" is really a batch AND:

⋀x P(x) = $P(x_1) \land P(x_2) \land P(x_3) \land ...$

Existential Quantifier, ∃

If we want to express that at least one person is in the club, or some people belong to the club, we are really saying that one person is, OR the next person is, OR…etc.

We are ORing propositions with different subjects, but the same predicate. If we represent the subjects (people) with a variable "**x**", then we can say some (at least one) **x** is **P**. We can use the existential quantifier symbol to represent that statement:

∃x P(x) Meaning: there exists an x, where x is P.

Some (at least one) person, is in the club.

We can memorise the symbol ∃ as an upside-down E for "Exists".

We could also use the OR symbol to express the same idea, some texts do. It reminds us that "there exists" is really a batch OR:

∨ x P(x) = $P(x_1) \vee P(x_2) \vee P(x_3) \vee ...$

Notation

In mathematics, we usually quantify or batch a summation (operator +) with the Σ notation:

$$\sum_{x=1}^{4} x^2 = 1+4+9+16$$

This is a very compact way to say "add up the numbers from 1 to 4, each squared".

However, this notation does have some limitations. One is that the range (x=1, 2, 3, 4) is limited to a contiguous stretch of whole numbers.

Note that the range is a Boolean. We can say the summation happens over the numbers that make the range true. A more flexible notation published by Gries & Schneider would write this as:

$(\Sigma x : \mathbb{Z} \mid 1 \leq x \leq 4 : x^2)$

At first glance it may look more awkward, but it allows us to be very specific and flexible about every aspect of the operation.

The components are:

1. Σ for the operator (+), & variable declared, x.

2. $\mathbb{Z}$ denotes the type of x ($\mathbb{Z}$ = integers, whole numbers).

3. The range ($1 \leq x \leq 4$) is a Boolean, x is either within it (range true), or outside it (range false). The operation is only performed where x is in the range (where the range is true).

4. Finally, the expression of x on which you are performing the operation (here, x squared).

This is the format for a quantification:

(operator variable : type | range : expression)

Now, say I have a box of 10 snakes. I may want to say something like:

($\forall s$: snake | $1 \leq s \leq 10$: $\neg P(s)$)

For all snakes, in the collection numbered 1 to 10, each snake is not poisonous. However, it is not quite correct to have snakes as numbers.

As the range can be any Boolean (true/false statement), there is an easier way:

($\forall s \mid B(s) : \neg P(s)$)

For all snakes, in the box, they are not P.

Where B stands for the predicate "is in the box", which allows us to deal with ranges that are not numerical.

We can leave the type unstated, if it is already obvious or known. If the range is unlimited, we can leave that space blank. We can then make further statements following from the above:

($\forall s \mid : B(s) \Rightarrow \neg P(s)$)

For all snakes, if they are in the box, then they are not P. The empty range signifies range ≡ true, that is, it includes all items of the same type as **s** (snakes).

For the operators ∀ and ∃, the last section of the notation, must contain a predicate (an expression that returns type Boolean). That is because these operators are really AND and OR, which can only operate on Boolean expressions.

We can also conclude:

¬ (∃s | B(s) : P(s))

It is not the case, that there exists some snake, in the box, that is poisonous. This transformation is a generalised case of De Morgan's law.

This notation is extremely flexible, and allows for complex ranges. For example:

(∀s | B(s) ∨ T(s) : ¬P(s))

All snakes in the box, or on the table, are not P.

Quantifications themselves can be Boolean propositions within larger arguments, for example:

¬ (∃s | B(s) : P(s)) ∧ (∀s | T(s) : G(s))

Chapter 5
The Logic of Sets

We have seen how predicate logic can allow us to categorise things. Now we need new operations and notation to allow us to perform logic with and between the categories themselves.

A set is a collection of distinct items. Within a set, no item can appear more than once.

We can describe a set in two ways. First as a list, for example, Jim, Fred, Jane, Sue. As a set, this would be written formally as:

S = { Jim, Fred, Jane, Sue }

The set **S** is defined as the collection of items within the curly brackets. This is called **set enumeration**, which could become very unwieldy with a large number of items.

The other way we can define a set is with a description, e.g. S = {people in this building}. This is called **set comprehension**.

An important thing to keep in mind is the **type** of object the set contains. The set {Jim, Fred, Jane, Sue} could be the set of people at a table, or the set of the four most common names in the building. They are very different types of things. Set comprehension can clarify this.

We can define sets using a format consistent with the notation already familiar to us:

$$S = \{\, x : \text{type} \mid R(x) : E(x) \,\}$$

Here a capital letter stands for a set, **S**. A dummy variable is declared, **x**. There is no operator, as we just want a list of items. The range **R** is Boolean as before, and the output to the list is an expression **E** of **x**. For example:

A sum $(\sum x : \mathbb{Z} \mid 1 \leq x \leq 4 : x^2) = 1+4+9+16$

A set $\{\, x : \mathbb{Z} \mid 1 \leq x \leq 4 : x^2 \,\} = \{\, 1, 4, 9, 16 \,\}$

$\{\, x : \text{people} \mid R(x) : \text{Name}(x) \,\} = \{\text{names of people in a room}\} = \{\text{Fred, Bill, Jane},\ldots\}$

$\{\, s : \text{snake} \mid B(s) : s \,\} = \{\text{Actual snakes in a box}\}$

Venn Diagrams

In order to visualise sets and their operations, we can use Venn diagrams. These provide an intuitive way to understand the mechanics of categories.

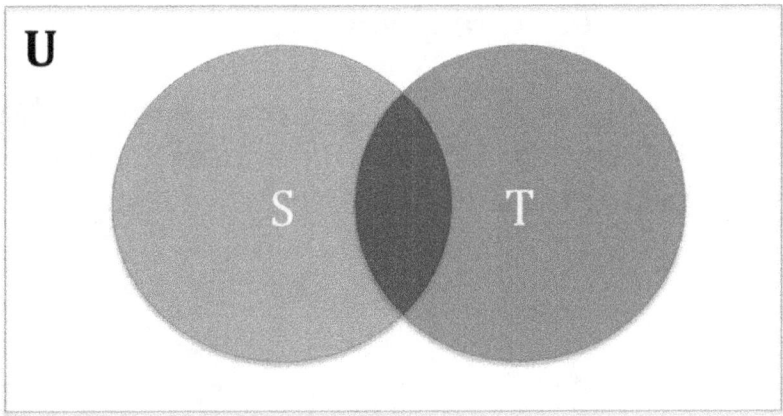

The rectangle represents the Universal set, containing all items of the type being discussed (the domain of discourse). **S** and **T** are sets, collections of unique items, which may have some items in common.

For example: U = {all people}, S = {Australians}, T = {all people who speak German}. Here indicating, some Australians speak German.

Set Operators

While the logical operators work on Booleans, and mathematical operators work on numbers, we also have operators that combine and transform sets in particular ways. These operators do, however, relate to the logical operators in a very direct way.

Also the symbols for the operators relate to the shapes of the relevant logical operators, to make the system intuitive and consistent.

Set Membership, ∈

Members of a set are also called **elements**. We can say that item **e** is an element of the set **S**:

$S = \{x \mid R(x) : E(x)\}$

$e \in S \equiv (\exists x \mid R(x) : e = E(x))$

The shape of the symbol reminds us of ≡, but the three lines emerge from a point. We can think of the element as a point within the set. It also reminds us of an E for "element". It returns a Boolean result (true or false), as an item either is or isn't an element of a set.

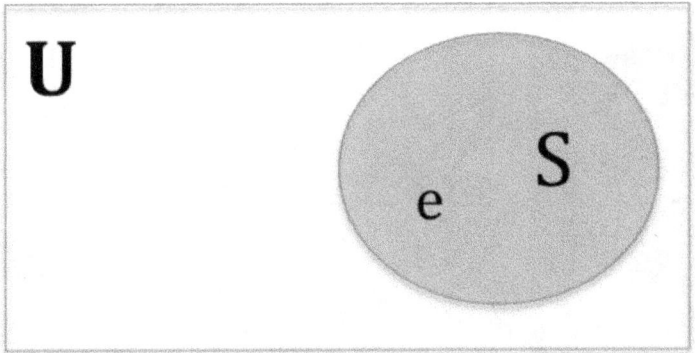

$e \in S$ showing element **e** within the set **S**.

Union, ∪

This operator creates a set that consists of all of the elements in the sets that it operates on:

S ∪ T = {all items that are in S **or** in T}

e ∈ {S ∪ T} ≡ (∃x | (x ∈ S) ∨ (x ∈ T) : e=x)

As we see above, logically ∪ relates to ∨ , and we see that the Union symbol reminds us of a rounded OR. This is why OR broadens a search. It is also easy to remember as it looks like a U for "Union".

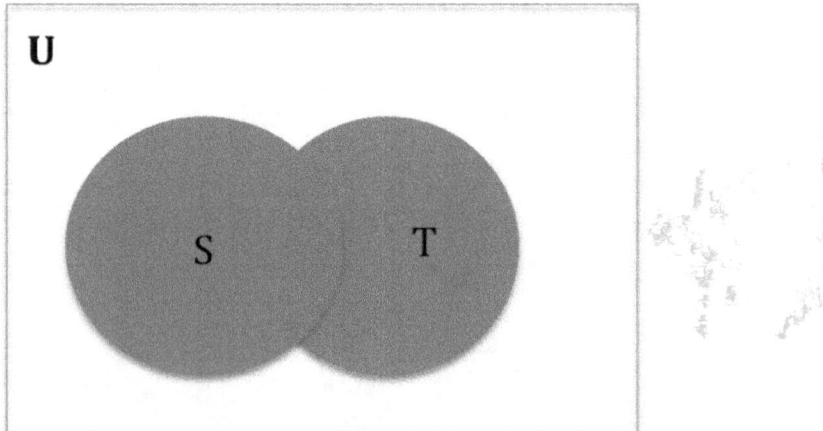

S ∪ T is the whole shaded area.

Intersection, ∩

This operator creates a set that consists of the elements that are shared by the sets that it operates on:

S ∩ T = {items that are in both S **and** in T}

e ∈ {S ∩ T} ≡ (∃x | (x ∈ S) ∧ (x ∈ T): e=x)

As we see above, logically ∩ relates to ∧, and we see that the Intersection symbol reminds us of a rounded AND. This is why AND narrows a search. It can be memorised as looking like the **n** in "Intersection".

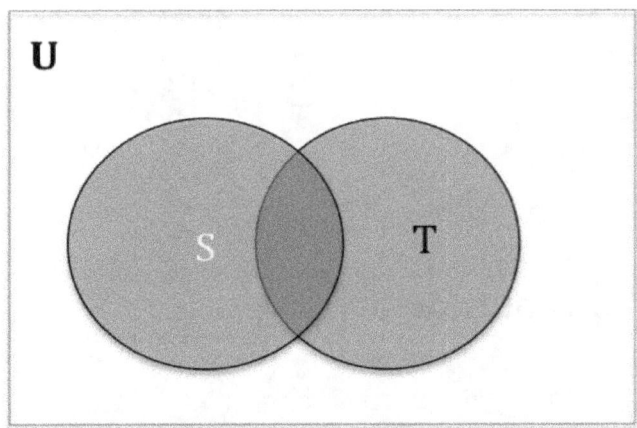

S ∩ T is the overlapping area.

Complement, ~

This operator creates a set that consists of the elements that are not in the set that it operates on:

~ S = {all items of **U not** in S}

e ∈ ~ S ≡ (∃x | (x ∈ **U**) ∧ ¬ (x ∈ S): e=x)

As we see above, logically ~ relates to ¬ , and we see that the complement symbol reminds us of – (negative) and ¬ (not).

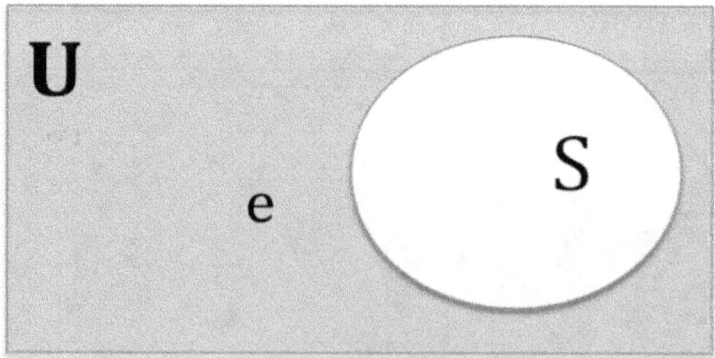

~ S is the area of **U** outside of **S**.

THE LOGIC OF SETS

Subset, ⊆

This operator returns a Boolean (true/false) when operating between sets. **S** is a subset of **T**:

$S \subseteq T \equiv$ (**if** an element is in S **then** it also must be in T)

$(S \subseteq T) \equiv (\forall x \,|\, : (x \in S) \Rightarrow (x \in T))$

As we see above, logically ⊆ relates to ⇒ . It also reminds us of ≤ , and the number of elements in **S** must be less than or equal to the number in **T**.

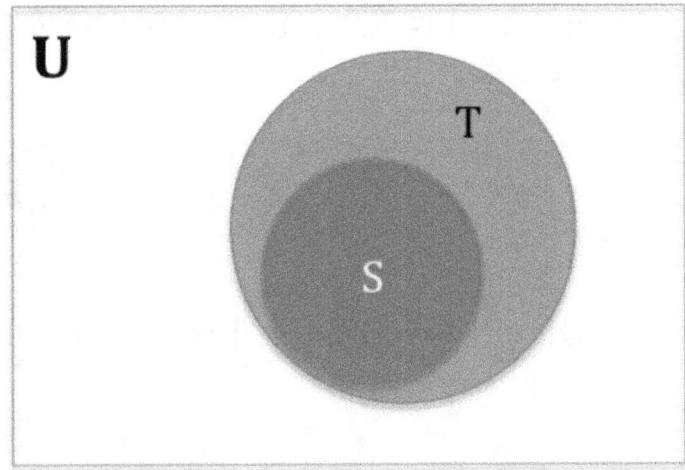

$S \subseteq T$ is true when **S** is within or equal to **T**.

Proper Subset, ⊂

The only difference between this and the previous operator is that the two sets cannot be the same. **S** is a proper subset of **T**:

S ⊂ T ≡ (an element being in S **implies** it also must be in T, **and** there exists an element in T that is not in S)

$$(S \subset T) \equiv (\forall x \mid : (x \in S) \Rightarrow (x \in T))$$
$$\land (\exists x \mid (x \in T) : \neg (x \in S))$$

As we see above, logically ⊂ relates to ⇒ . It also reminds us of < , and the number of elements in **S** must be less than that in **T**, except if they are infinite sets.

For example, the set of even numbers is a proper subset of the set of integers, but they are both still infinite.

(S ⊂ T) is true when **S** is within **T**, but not equal to **T**.

Set Equality, =

Two sets are equal if all of the elements in each are also in the other. This operator returns a Boolean, true if the sets are the same.

$S=T \equiv$ (all elements in each are in the other)

$S=T \equiv (\forall x \mid : ((x \in S) \Rightarrow (x \in T)) \wedge ((x \in T) \Rightarrow (x \in S)))$

$\equiv (\forall x \mid : (x \in S) \equiv (x \in T))$

We can see logically, this relates to the biconditional ($\equiv$, $\Leftrightarrow$). We can also say that each set is a subset of the other.

An example could be {people in the building whose names start with B} and {people in the building whose birthdays are this week}, although the sets are comprehended differently, and are not necessarily the same. Just by coincidence, they may both enumerate the same set of people, and the sets would be the same.

Set Difference, -

This operator creates a set that consists of the elements that are in the first but without those that are in the second.

S – T = {items in S **and not** in T}

e ∈ {S – T} ≡ (∃x | (x ∈ S) ∧ ¬ (x ∈ T): e=x)

As we see above, logically – relates to ∧¬ , and reminds us of the familiar operation of subtraction. However, the elements removed from **S** are only the intersection of **S** and **T**.

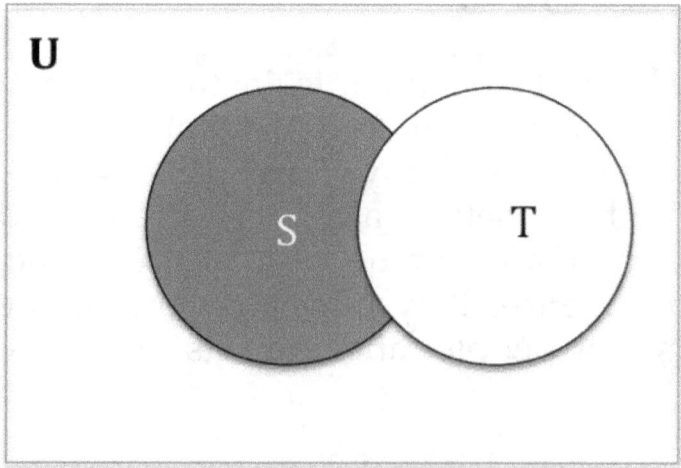

(S – T) is the shaded area.

Special Sets

U, The Universal Set, Domain of Discourse, a type.
$=$
{ x : type | : x }

∅, the Empty Set. ∅ = { }
$=$
{ x | false : x }

{e}, Singleton Set. Set with one element.
$=$
{ x | x=e : x }

#S, Cardinality. Number of elements in a set.
$=$
($\sum$x | x ∈ S : 1)

Sets have a type, and their elements have a type.

The type of the set {1,2,9} is "set($\mathbb{Z}$)".

A type is the set of all values that type can have:
$\mathbb{B}$ = {true, false} (Boolean)
$\mathbb{Z}^+$ = {1,2,3,...} (positive integer)

Chapter 6
Logic and Uncertainty

People tend to think of logic and science as dealing with, or at least aiming for, certainty. In fact the opposite is more often true. They are more about how to deal with uncertainty.

Testing and Detection

Nothing we know about the world is certain. We can be very close to certain about some things, and very sure about the unlikelihood of others, as well as any position in between. In order to understand what is going on around us, or make good plans, we need be able to judge the likelihood of various possible interpretations we may have in mind.

If we hear a noise, we may think of a tiger. However, we need to consider the likelihood of being correct, which will depend very much on our location. Acting on the wrong assumption may put us into increased danger. Such decisions are made daily, and often have serious consequences.

Everything we know about the world is the result of some form of detection. Whether directly by our senses, or via instruments, data is acquired and interpreted. This is the process of detection, and it is never perfect.

In the real world, the raw data available to us is a mix of random or irrelevant "noise", and relevant patterns. Data cannot be considered evidence until it is extracted from the noise, ie detected.

Detection is a test for a particular pattern in the raw data, and comes with two types of errors.

Type I, false positive: in which the test detects the pattern when it is not really there.

Type II, false negative: in which the pattern is present but is not detected.

Quite often, we can obtain or deduce fairly accurate estimates of the error rates. From this, we can make informed decisions based on likelihoods, or **probabilities**.

Once we can work out how likely or unlikely things are, we are much more able to assemble the jigsaw puzzle of information that reveals the world to us.

Probability

First, we need to know what a probability is, and how it is measured. The probability of a proposition being true, or an event occurring is simply the number of times it is expected to occur out of the maximum number of times it could occur. That is, the proportion of times it is true.

If we let the proposition **r** be "it is raining". The probability of **r**, that it is raining, is the number of times we expect to find it raining out of the number of times we check for rain:

$p(r) = \#r / \#(r \vee \neg r)$

More generally, we could say:

$p(R(t)) = (\sum t \mid R(t): 1\,) / (\sum t \mid R(t) \vee \neg R(t): 1\,)$

$\qquad = (\sum t \mid R(t): 1\,) / (\sum t \mid : 1\,)$

The probability of a statement **R(t)** being true is the number of trials, tests, times, intervals, or detection opportunities **t** in which the predicate **R** is expected to be true, out of the total of **t**. In other words, the fraction of **t** in which **R(t)** would be true. Note: (r∨¬r) ≡ true.

For example, if we toss a coin, we expect that it will land heads about half of the times it is tossed. The probability of tossing heads is one half, or 0.5. If we toss it 10 times we expect 5 heads on average:
$p = 5/10 = 1/2$

If a trick coin had two heads, we would expect heads every time, the maximum possible:
$p = 10/10 = 1$

If the coin had two tails, we would expect heads no times, the minimum possible:
$p = 0/10 = 0$

From this we can see that a probability has to be a number between 0 and 1.

We write for an event, the proposition **s** (the event happens), and the probability of **s** being true is $p(s)$, where $0 \le p(s) \le 1$. Also:

$(p(s) = 0) \equiv$ (s is impossible, always false)

$(p(s) = 1) \equiv (s \text{ is certain, always true})$

$p(s \lor \neg s) = p(s) + p(\neg s) = 1$

By the law of the excluded middle either **s** or not-**s** has to be true, so the probability that one or the other is true is 1, or certainty. If the probability of tossing heads is 0.5, then the probability of not-heads (tails) is 0.5, and the probability of either heads or tails is 1.

In dice, if we roll a die, there are 6 equal possibilities. The probability of rolling a 3 is 1/6, and the probability of not rolling a 3 is 5/6, which is the probability of any of the 5 other possibilities happening. The total probability of every possibility in a system has to add up to 1. If **d** is the number resulting from a roll of a single die, then:

$p(d=3) = 1/6$

$p(\neg(d=3))$

$= p((d=1) \lor (d=2) \lor (d=4) \lor (d=5) \lor (d=6))$

$= 5/6$

Simple and Compound Events

A simple event is an event that can only happen in one way, for example, the toss of a single coin, or the roll of a single die.

A compound event is one that can happen in more than one way, for example, if we toss two coins, there is only one way to get two heads. This is a simple event. However, there are two ways to toss a head and a tail (h,t) or (t,h) even though we do not distinguish between the two coins, and they land at the same time.

If we did not know this, we might assume that the probability of two heads, two tails, and mixed, could all have the same probability of 1/3 each.

However, the set of all possible results is actually:

S={hh, ht, th, tt} each result with equal probability of ¼.

So we can see that the coins will land two heads a quarter of the time, two tails a quarter of the time, but mixed half the time.

This can be verified experimentally at home.

Sets in Probability

Sets are useful in visualising compound events, and complex event spaces.

We can think of a simple event as an element **x** with a probability **p(x)**. We can think of a compound event as a set of elements where x∈X. We can shorthand the probability of an event from the set happening, as the probability of the set X, **p(X)**, which will be the sum of the probabilities of each of its elements.

In the example of two coins, there are several ways we can define compound events. We can define one where the two coins are the same {hh, tt}, or one with at least one tail {ht, th, tt}. The probability of each compound event will depend on how many simple events it has.

Using the above example:

V = {both same} = {hh, tt} ; $p(V) = 2/4 = ½$

T = {some tails} = {ht, th, tt} ; $p(T) = 3/4$

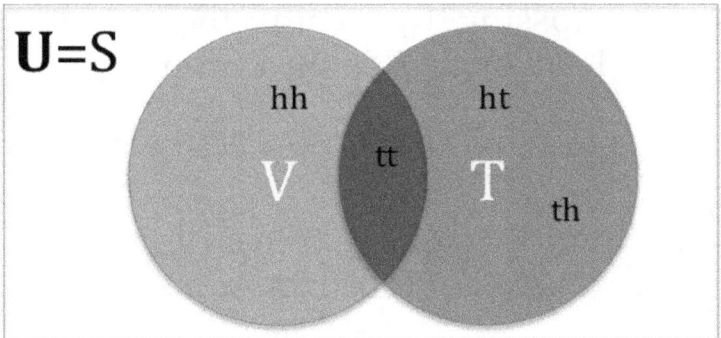

A Venn diagram illustrating the events.

The Universal Set or domain is often called the Sample Space **S** in probability, and contains all possible simple events. We will use this convention when dealing with probabilities.

Set operators allow us to calculate the various combinations of events and their probabilities in a complex event space.

For example, the event of tossing two coins the same, **and** containing at least one tail is the intersection of V and T:

$V \cap T = \{tt\}$; $p(V \cap T) = 1/4$

The event of tossing two coins the same, **or** containing at least one tail is the union:

$V \cup T$ = {hh, ht, th, tt} = S ; $p(V \cup T) = 4/4 = 1$

This example also demonstrates how AND narrows a search, and OR broadens a search.

Obviously, the probability of a whole sample space has to be 1, as by definition it contains all of the possible outcomes. In other words, it is certain that one of the possible outcomes will happen. Even if the result of "nothing happens" were possible, that would still have to be an outcome with some probability, and part of the sample space:

S = {all possible outcomes} , $p(S) = 1$

Conditional Probabilities

Sometimes we need to know the probability of an event under different conditions. For instance the probability of road accidents in wet weather, compared to dry.

This concept is at the heart of the empirical evidence in science, and the reason for controlled experiments.

Now say we want to build a detector that will tell us when particular kinds of things happen. We specify a definition, such that the predicate **P** means that event **x** is an event of interest. We will call this set **A**:

$A = \{ x \mid P(x) : x \}$

We now define a set **B** as the set of events that actually trigger the alert that the detector has found a match. We may specify **P'** to mean the detector has matched an event to a signature that should describe the events of interest:

$B = \{ x \mid P'(x) : x \}$

In a perfect world, A=B. So x∈A ⇒ x∈B, in shorthand A⇒B. Also B⇒A. If an **A** event happens the **B** alarm goes off, and if the alarm goes off it is a real event **A**. We will use the notation *p*(**B** | **A**), to signify the probability of **B** if **A** is true, i.e. the probability of "**B**, given **A**":

(A⇒B) ≡ (p(B | A)=1)

A implies **B**, is the same as saying that if **A** is true then **B** is certain.

However in the real world detection involves errors. *p*(**B** | **A**) is the probability of getting an alert when an event of interest happens. It is a measure of the sensitivity of the detection, and usually the most important consideration.

If we buy a detector, or go for a medical test, we are most concerned with the ability of the test to detect what we are looking for. We accept that the sensitivity of the detection may mean an occasional false positive result. A positive result can always be re-tested to make sure. Failing to detect a condition, a false negative, usually has more serious consequences.

Examples include: a medical test. A false positive may be a worry, but will be cleared in further tests. A false negative could delay treatment until it becomes too late. A fire alarm may be annoying if it is set off by steam, but if it fails in a fire, it could be deadly.

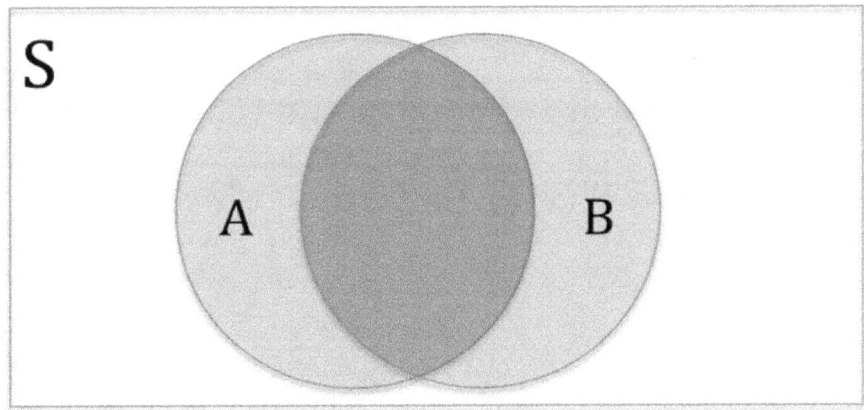

First we want to know the probability of getting an alarm if there is an event:

$p(B \mid A) = p(A \cap B)/p(A)$

This is the proportion of **A** events that trigger an alarm **B**, or the proportion of the set **A** that is shared with set **B**. In other words, the fraction of **A** covered by **B**.

False negatives have probability $p(A-B)$. False positives have probability $p(B-A)$. Errors are represented in the areas of each set **A** and **B**, where they do not overlap.

In the intersection:

$A \cap B$

= {events that are of interest **and** trigger alert}

= $\{\, x \mid P(x) \wedge P'(x) : x \,\}$ = {True Positives}

Bayes' Theorem

Sometimes we need to know $p(B|A)$, but we can only directly know $p(A|B)$. These are very different things. Like implication, inference cannot simply be reversed.

As we saw in the previous section:

$$p(B|A) = p(A \cap B)/p(A)$$

But we can also see that probability of there being a genuine event if the alarm goes off will be:

$$p(A|B) = p(A \cap B)/p(B)$$

This is the proportion of **B** that is shared with **A**. As the intersection of **A** and **B** gives us a term in common, we can re-organise this:

$$p(A \cap B) = p(B|A)\,p(A) = p(A|B)\,p(B)$$

Now we can relate the two conditionals:

$$p(B|A) = p(A|B)\,p(B)/\,p(A)$$

Chapter 7
Putting Logic to Use

Once familiar with the basic concepts of logic and probability, we can start to make sense of situations that most people find confusing.

Some good sources for examples of suspect logic are advertising, discussions on the Internet, political arguments in the news.

For the more extreme examples look up popular sites for conspiracy theories, or similar arguments against well-established science, like the anti-vaccination scares, although these tend to rely as much on basic ignorance in their premises as on faulty logic.

This highlights the fact that logic alone is not enough. We also need to have a basic knowledge of the world in order to provide credible foundations for any argument. This includes the rudiments of scientific literacy.

The best way to understand the uses of logic is through examples. All of the following examples are solved using the concepts already explained in this book. For this reason, some of the steps are not spelled out. For example, the detection puzzle has three probabilities given, but a further three are deduced. Hint:

$p(X) + p(\sim X) = 1$

Propositional Puzzle

"We interviewed the three suspects, Blake, Jeeves, & Lee. Blake insisted that Jeeves is guilty, while Jeeves says Blake is guilty only if Lee is, but then Blake says Lee is innocent. Lee, of course, claims he is innocent, and it was one of the other two who did it."

Is there a consistent story in this, and if so, are we able to identify the guilty parties?

Some may guess correctly, but will find it very difficult to explain clearly why.

First we need to identify the minimum number of propositions:

b = Blake is guilty
j = Jeeves is guilty
l = Lee is guilty

These are the only propositions in the argument.

Next we need to identify the logic:

Blake says "**j** and not **l**" (j ∧ ⌐l)
Jeeves says "**b** only if **l**" (⌐l ⇒ ⌐b) and so (b ⇒ l)

Lee says "not **l** and either **b** or **j**" (⌐l ∧ (b ∨ j))

We can then apply a truth table:

b	j	l	(j ∧ ⌐l)	(b ⇒ l)	(⌐l ∧ (b ∨ j))
t	t	t	f	t	f
t	t	f	t	f	t
t	f	t	f	t	f
t	f	f	f	f	t
f	t	t	f	t	f
f	**t**	**f**	**t**	**t**	**t**
f	f	t	f	t	f
f	f	f	f	t	f

We can now see that the only way all three statements can be true is if Jeeves is guilty, and the other two are not.

Predicate Puzzle

If we have a statement from a previous example "all snakes in the box are not poisonous":

$$(\forall s \mid B(s) : \neg P(s)) \equiv \neg (\exists s \mid B(s) : P(s))$$

We can intuitively deduce that "there is not a snake in the box that is poisonous". So much so, that we may unconsciously use the two statements interchangeably in an argument. Indeed, as shown above, the two statements are logically equivalent.

We can reason that if all of the snakes are non-**P**, then there cannot be one that is **P**. It makes sense

As mentioned earlier, this is a generalised De Morgan's rule. So how does it work? Remembering the rule:

$$(\neg p \wedge \neg q) \equiv \neg (p \vee q)$$

In its basic form, we have two propositions. All we need to do is extend it to many propositions.

$(\forall s \mid B(s) : \neg P(s))$
$=$
$\neg P(s_1) \wedge \neg P(s_2) \wedge \neg P(s_3) \wedge \ldots \neg P(s_n)$:Definition of $\forall$
$=$
$\neg (P(s_1) \vee P(s_2) \vee P(s_3) \vee \ldots P(s_n))$:De Morgan
$=$
$\neg (\exists s \mid B(s) : P(s))$:Definition of $\exists$

Of course it also works for the other half of the rule:

$(\neg p \vee \neg q) \equiv \neg (p \wedge q)$

Is generalised to:

$(\exists s \mid B(s) : \neg P(s)) \equiv \neg (\forall s \mid B(s) : P(s))$

If we say "there exists a snake in the box that is not **P**", we also can say "it is not the case that all of the snakes in the box are **P**".

We can note here that some people say "all **s** are not **P**", when they really mean "not all **s** are **P**". In logic these are very different statements:

$\forall s \neg P(s)$ is not the same as $\neg \forall s P(s)$

Quantification Puzzle

How can we express something seemingly simple like **y** is an even number?

We know that even numbers are divisible by 2. Assuming we are only looking at positive whole numbers, we can construct a test for evenness.

If **y** is even, then there must be a whole number (integer) **x** that can be doubled to produce **y**:

$(\exists x : \mathbb{Z}^+ \mid : 2x = y)$

If there exists a whole number **x**, that can be multiplied by 2 to equal **y**, then the quantification is true, and **y** is even. If the quantification is false, there is no **x** that can be doubled to make **y**, and **y** is not even.

We could insert a range $(\exists x: \mathbb{Z}^+ \mid 0 \leq x \leq y : 2x = y)$, which may be useful if you are describing it as an algorithm to someone, and want to steer them to the relevant interval, to avoid potentially wasting time.

Logic and Probability Puzzle

This example is from a real discussion online. A nurse was commenting on a discussion on the efficacy of a vaccine.

Her analysis was that in her hospital there were about 200 cases of the illness that year, all patients were asked about their vaccination status. It turned out that there was about an equal number of immunised and non-immunised patients presenting as sick. Therefore, she concluded, the vaccine must make no difference.

Further to this, she commented that it seemed like a commercial conspiracy that over 90% of the population in her town were conned into receiving an ineffective treatment.

A few people commented that it "proved" their position. A few commented that her information must be wrong. Of at least a dozen educated people, nobody questioned her logic.

In fact there are a number of errors in the argument. The first is appeal to authority.

Many assumed that a nurse must be a reliable commentator.

Secondly, a reversal of the inference: The evidence was the rate of vaccination, given that people were sick. The conclusion was about the rate of sickness, given that people were vaccinated.

Thirdly, the nurse was only considering her limited view of the hospital, and did not consider the whole population in question.

By her own numbers, over 90% of the general population was vaccinated, which means that there are about 10 times as many vaccinated people as non-vaccinated. As the number of people getting sick in each group was equal, the proportion of sick for the non-vaccinated population is in fact 10 times as large. So the probability of the unvaccinated getting sick is ten times as great.

This explanation still met with some confused replies, until I drew a diagram, which ended the debate. To make the calculation simple, I have used round numbers, but it is the relative proportions that determine the outcome, so changing the actual numbers would make no difference to the ten-to-one result.

The rectangle represents the hospital, while the ovals represent the population of the town, split into immunised and non-immunised.

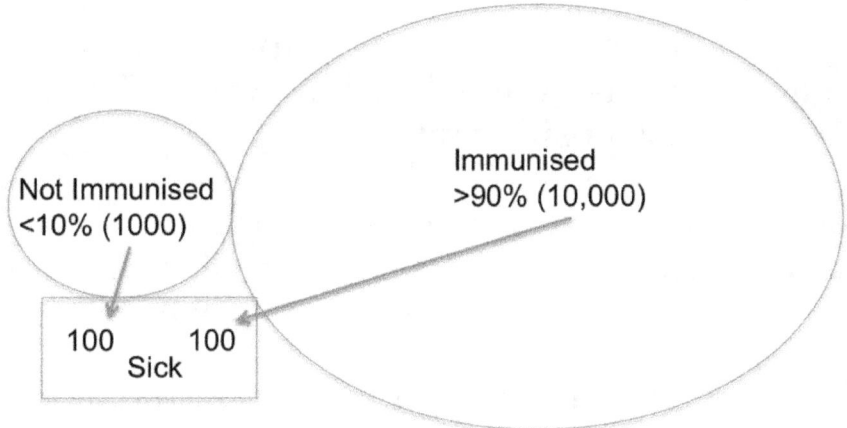

$p(I \mid S) = 0.5$
(chance of sick patient having been immunised 50%)

$p(S \mid {\sim}I) = 100/1000 = 0.1$
(chance of non-immunised person getting sick 10%)

$p(S \mid I) = 100/10{,}000 = 0.01$
(chance of immunised person getting sick 1%)

Events and Detection Puzzle

We buy a detector to alert when a certain kind of event happens. The manufacturer provides some marketing material that boasts that it will detect 99% of these events.

Suitably impressed, but aware of the two types of error, we agree that the implied 1% false negative rate is acceptable, but want to know what the false positive rate is.

The manufacturer provides specifications that show a 10% false positive rate. So far, we are satisfied that the detector is what we want.

Now our operators want to know, of the alerts they receive, how many will be false positives. Do we have enough information to answer? Not yet. We need to know the expected rate of events of interest in our system. We do some research and find that 5% of all events will be of the type we want to trigger alerts.

Most people asked to guess the percentage of alerts that will be false, will answer between 10% and 30%. We will work it out.

So far, we have:
$p(B \mid A) = 0.99$
$p(B \mid \sim A) = 0.1$
$p(A) = 0.05$

Now we can work out the probabilities for all of the possible outcomes, which are: true positive, false negative, false positive, and true negative.

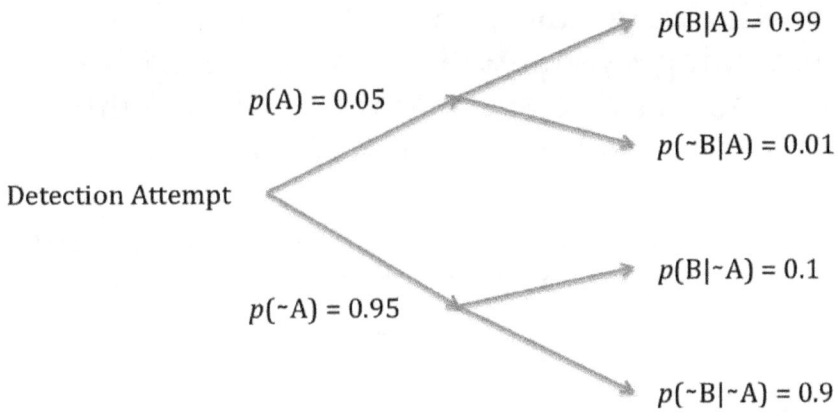

$p(A \cap B) = p(A)\, p(B \mid A) = .05 \times .99 = .0495$ (p of alert **and** it is true)

$p(B) = (.05 \times .99) + (.95 \times .1) = 0.1445$ (p of an alert)

$p(A \mid B) = p(A \cap B)/p(B) = 0.343$ (p of alert being true)

$p(\sim A \mid B) = 1 - p(A \mid B) = 0.657$ (p of alert being false)

Using Bayes' rule, we can see that two thirds, or about 66%, of all alerts will be false positives. This comes as a surprise to most, and reminds us once again that probabilities are often not intuitively obvious. However, it should not be surprising if we think in terms of proportions.

We knew that 5% of all events are the type we want to alert on. We know that nearly all of them will trigger alerts. We also know that 10% of remaining events will trigger false positives. So it is not hard to see that there will be almost twice as many false alerts as real ones.

It is easier to visualise with a Venn diagram in which area is made proportional to probability.

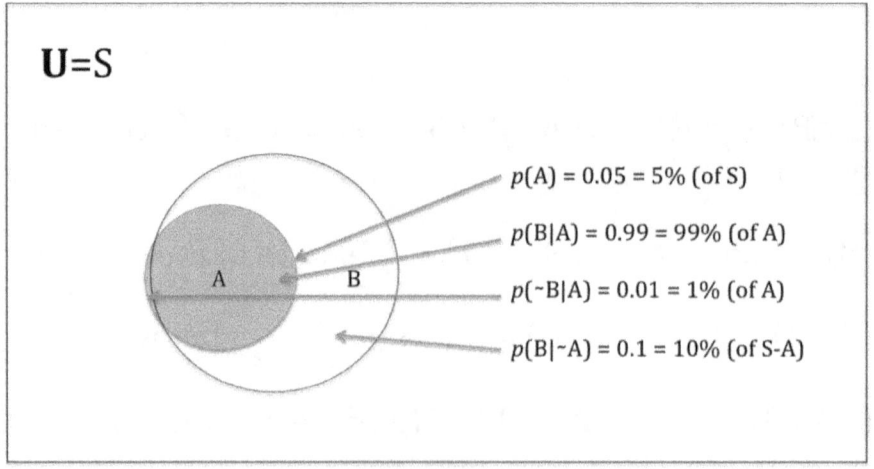

We may ask why we don't reduce the sensitivity, in order to reduce false alarms.

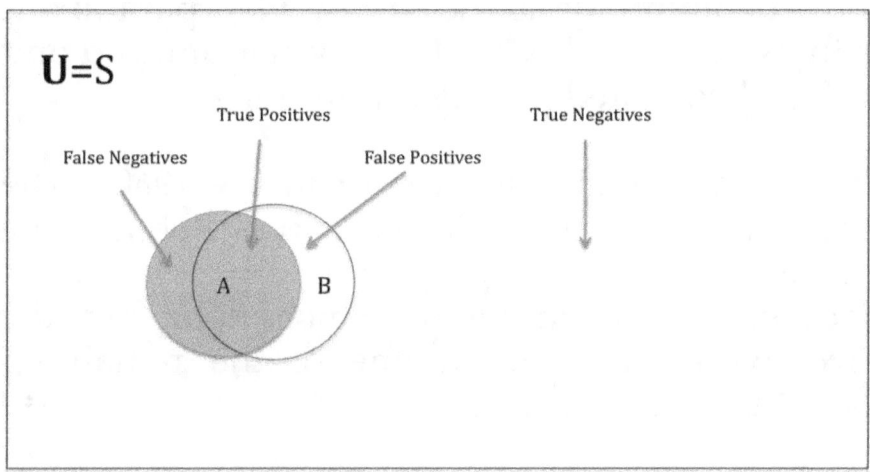

In general, if we try to reduce false positives we will increase false negatives by a much larger factor. In this case, reducing false positives to about 1/3 of previous levels, has increased false negatives to around 30 times previous levels. This would be unacceptable for most detectors or tests. Of course, in practice, we tune detectors to align **A** and **B** as much as possible, to reduce both types of errors.

(Puzzle adapted from a lecture in MIT course "Probabilistic Systems Analysis and Applied Probability", iTunesU)

Classic non-Paradox of Probability

Known commonly as the Monty Hall paradox, it really isn't a paradox at all. Yet it has amused and confused educated people for decades.

Monty was a game show host in the 1960s. The game involved three closed doors. Behind one door was a prize. The guest would pick a door at random. Monty, knowing which door hid the prize, would then open one of the remaining doors that had no prize.

The guest was left with the guarantee that the prize was either behind the door chosen at random, or behind the one other closed door.

Monty now asks the guest to either stick with their first choice, or switch to the other closed door. Most people intuitively stick.

However, statistically, it is significantly better to switch, and this was verified experimentally on the TV show "MythBusters".

Most people cannot immediately see why this is true, and once again we are confronted by our blind spot when it comes to probabilities.

We have no trouble seeing that the chance of picking the prize initially is 1 in 3, $p=(1/3)$.

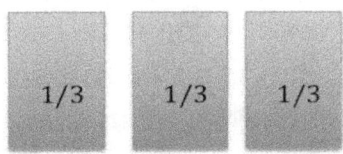

Now say we pick door 1.

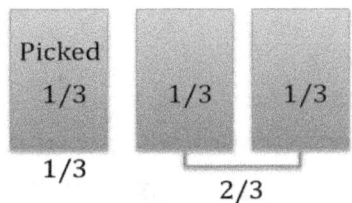

Now Monty opens door 3, to reveal no prize.

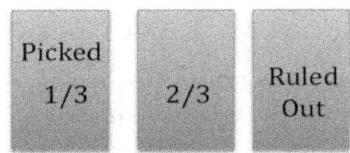

We are now twice as likely to find the prize behind door 2 as door 1. At this point, many will still not be convinced.

Perhaps it is better to think of this as the least case of a more general problem.

Say we had a large number of doors, for example a thousand. You pick one at random, $p=(1/1000)$, very unlikely to pick the prize.

Then Monty reveals 998 empty doors, all but yours and 1 other are left. The prize is guaranteed to be either behind the one you picked at 1 in 1000 odds, or the other one that he has virtually told you the prize is behind $p=(999/1000)$. You shouldn't have any trouble deciding to switch now.

A completely equivalent example: I hold a shuffled deck of cards fanned out so that I can see them, and you can't. Now I tell you that one of us will definitely pick the Ace of Spades. You win a prize if you hold that card. I get you to pick one at random and lay it face down. Now I pick one deliberately, still viewing the faces, then lay it face down. I know what both cards are, and you don't.

You can now choose. Do you stick with your pick, or swap with mine?

Rolling Dice

We have seen that if you roll a die, it has 6 possible outcomes with a sample space of the set $S= \{1,2,3,4,5,6\} = \{x \mid 1 \leq x \leq 6 : x\}$. If the die is fair, each outcome **d** has the same probability $p(d=x)=1/6$.

However, many games involve rolling two dice and adding the two results together. The possible outcomes are $S=\{2,3,4,5,6,7,8,9,10,11,12\} = \{x \mid 2 \leq x \leq 12 : x\}$.

If we know the dice are fair, many would expect that the probability of each of the 11 outcomes is the same, $p(d_1+d_2=x)=1/11$. This is not correct.

For each of the six sides of one die, there are six possibilities for the other. This means the total number of possible combinations is $6 \times 6 = 36$. The probability of rolling a 2 is $p(d_1=1 \wedge d_2=1) = 1/36$. $p(d_1=1) = 1/6$ and $p(d_2=1) = 1/6$. The probability of both die landing on 1 is the probability of each multiplied. So far, so good.

Now when we look at the probability of rolling a 3 is $p((d_1=1 \wedge d_2=2) \vee (d_1=2 \wedge d_2=1))$, even though we do not distinguish between the two dice. This is a compound event, as there are two ways to get a 3. The probability is therefore $(1/6 \times 1/6) + (1/6 \times 1/6) = 1/36 + 1/36 = 2/36$

So rolling a 3 is twice as likely as rolling a two. How likely is a 4?

$p((d_1=1 \land d_2=3) \lor (d_1=2 \land d_2=2) \lor (d_1=3 \land d_2=1))$

$= (1/6 \times 1/6) + (1/6 \times 1/6) + (1/6 \times 1/6) = 3/36$

Observe how ∧ multiplies probabilities, while ∨ adds them. For the same reason that AND narrows a search, while OR broadens it.

We can soon work out that a 5 can happen 4 ways, and a 6 can happen 5 ways. How many ways to make a 12? Not 11, but only 1, (6+6).

$4 = 1+3 = 2+2 = 3+1$ (3 ways, so $p=3/36$)

$6 = 1+5 = 2+4 = 3+3 = 4+2 = 5+1$ (5 ways, $p=5/36$)

We can see that a 6 is five times as likely as a 2. We can plot this on a column graph.

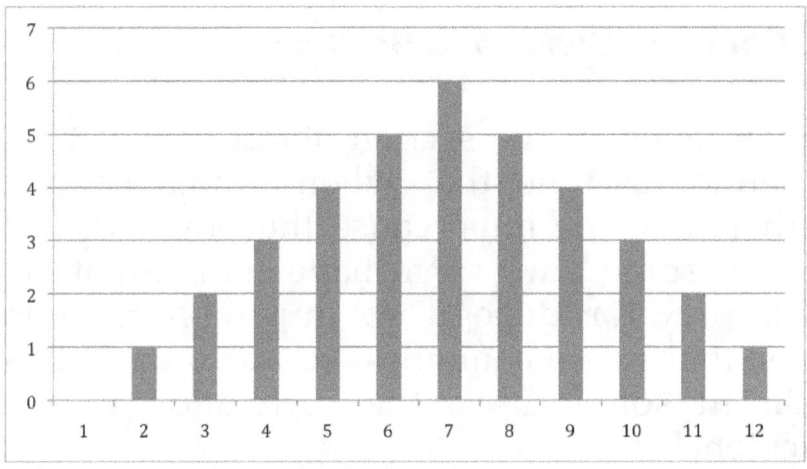

Here we see how many ways each result can be obtained, and that the most likely outcome is 7, with a probability of $6/36 = 1/6$. It is six times more likely than a 2, and twice as likely as a 4.

This can be verified experimentally at home, just keep rolling a pair of dice and count how many times each number comes up.

Common Sense and Science

As mentioned in the section about implication, we cannot assume that when things happen together that one must cause the other. In fact there are some things that have causes that are so complex, or depend so much on random events, that it is meaningless to attribute causes at all. In some cases, the correlation is just coincidence.

Human nature makes us look for patterns. Noticing patterns allows us to understand the world, but it can also lead to superstitious thinking.

Unfortunately, perhaps due to lawyers and accountants and litigious clients, there has been an increasing faith in the myth that everything can be ascribed a cause, and there is someone to blame in every situation.

We certainly need processes for quality improvement, and accountability for negligence, but we seem to have lost sight of the balance. Every action contains risk, and often risk is a necessary price for getting things done. If we penalise those who take reasonable risks, we stifle innovation, and paralyse progress.

If we accept that we can blame others for all of life's misfortunes, we lose the incentive to take responsibility for ourselves. Worse, we can lose the ability to accept our situation and move on. It is neither logical, nor healthy for the individual or society.

Perhaps the worst manifestations of the blame culture, are the proliferation of anti-science philosophies, and so-called conspiracy theories. While we can fault some of them on logical grounds, most of the strident speculation is based on false assumptions and basic ignorance about how the world works. This makes it possible to blame any of the world's problems on science, or any chosen organisation or group.

Another problem is the loose use of the word "theory". There is a yawning gulf between the idle speculations of the conspiracy buffs, and the stringent levels of evidence and consensus required to gain "theory" status in the scientific sense. If I hear someone say "only a theory", I know they are not speaking from a scientific perspective.

There does seem to be a bit of a crisis in critical thinking going on. We have never had so much tolerance for dissent and alternative ideas. This could be a good thing, if it leads to innovation and creativity. However, it just as often seems to lead to a complete failure of common sense.

Outlandish fairytales are accepted without regard to the credibility of sources, or the tried and tested principle of Occam's Razor. This principle states that the hypothesis requiring the least elaboration and credulity is usually the correct one. The simplest explanations are the most likely ones.

Education is the key. Schools need to teach the basics of critical thinking, along with a broad general knowledge, and the skill of finding reliable source material. It needs to be taught in an interesting way, so that it does not just incite rebellion.

People believe ridiculous things because they seem more exciting, easier to learn, or more convenient. This leaves the door open for fads, cults, and extremism of various kinds. As the late great Carl Sagan used to say "gullibility kills".

The first step is to teach children to be curious about the world. As they get older, encourage them to question, and learn how to sort reliable information from rubbish.

Part of the problem is the assumption that belief is something to be committed to. It is put forward as a virtue. There is an expectation that you either commit to a proposition or its negation. There is sometimes the idea that to be undecided, or uninterested in believing is not a good position to be in.

If we look more closely, belief is nothing more than an emotional attachment to a false sense of certainty. Once free of the habit of believing, we are free to see things more clearly, without the filters of preconception, and without the burden of shoring up cherished dogmas against reality.

Abandoning belief does not, however, preclude faith or spirituality, if we understand this to be a personal relationship with the Universe. Whatever that may mean to the individual, and whatever our experience of the sacred, it has nothing to do with making or believing unprovable statements.

Many assume that science is about collecting facts for people to believe. Actually, it is about how to decide on the best explanations and descriptions of reality, without committing to certainty.

Some have taken this to mean that there is no reality, so it makes no difference what people believe. The absurdity of this claim is easy to see through. The fact that we can detect consistent patterns in the environment, through our senses or instruments is enough. The simplest explanation is that there is an underlying structure generating or reflecting the patterns, and that is the real world.

One problem is that people do not know who to trust. They are unable to assess the credibility of sources. Again, we need to use common sense. Which is more likely to be true, the opinion of a large number of people who have done years of relevant study, have reputations to maintain, and have reviewed each others work, or the ideas of politically motivated non-experts who cite anecdote and selective data as their evidence?

Governments and corporations have to accept much responsibility for public distrust. Cover-ups of public dangers, like BSE, SARS, and smoking, by Britain, China, and the tobacco companies respectively, have shaken public confidence in the "establishment", but even worse, such dishonesty has created a disregard for evidence-based knowledge itself.

If we are to act capably and successfully, dealing with the challenges of the future, we need to be able to identify the best sources information, and be able to reason and plan. This requires the elements of logic, reliable common knowledge, and a basic high-school level of scientific literacy, for a majority of the population. It does not need to be highly technical coverage, but at least an awareness of good and bad arguments.

For technical, legal, and scientific careers, a familiarity with formal logic is essential, at least to the basic level covered in this book.

The best practical guide to critical thinking is science. It is a reliable source of information because its method is cautious and logical.

The scientific method starts with observation. A pattern is detected. The observation is retested, refined, and checked by numerous others trained in the field. This may involve experiments conducted to determine how various conditions affect a measurement. Once thoroughly verified as a real signal, an attempt is made to explain it by existing theories.

If the observation cannot be explained, new hypotheses are put forward. The simplest and most likely are chosen to test. By logical deduction, the predictions of the competing hypotheses are determined. Experiments are designed to test whether the results of each hypothesis are supported by experimental results.

An important part of the process is that experiments are always designed to disprove the hypothesis. We can never prove that the hypothesis is true. We can only show that it is supported by evidence if we are able to test for, but not able to produce, a result that disproves it.

A speculation that cannot be tested in this way can never become a theory. Most pseudoscience is based on speculations that can never be proven true or false because they rely on some component that cannot be detected or measured in some way.

If a hypothesis adds a significant piece to the body of knowledge, and it is well tried, tested, and widely accepted, it may be elevated to the status of a theory.

It is almost paradoxical that by embracing uncertainty, science comes closer to practical certainty than any other method of gaining knowledge.

Further Reading

Here are some useful books. Most are available in electronic format, or second hand online, and very affordable for students.

For a good general introduction to logic:

Introducing Logic, D. Cryan, S. Shatil, B. Mayblin

More advanced, formal, and a little dated:

Introduction to Logic, A. Tarski

Mathematical focus but impeccable notation:

A Logical Approach to Discrete Math, D. Gries, F. Schneider

Eloquent critique of irrationality:

The Demon-Haunted World, Carl Sagan

Exposé of illogical argument styles:

Believing Bullshit, S. Law

Belief-free spirituality:

Waking Up, S. Harris

About the Author

I have been interested in logic and science since I first watched Mr Spock on Star Trek from about the age of six. I continued my interest through my school years.

After a couple of years studying physics and maths at ANU, some time in the British Army Royal Engineers, a couple of years back at ANU studying psychology, logic, statistics, and linguistics, I finally discovered the then rather new industry of personal computing in the mid 1980s. I have been working in the IT industry ever since.

I have been working at IBM since 1997, where I have been able to explore various areas of technical interest. I am currently working in digital forensics.

I am a member of the Australasian Society for Human Biology, the National Space Society, and Mensa, and have published a variety of articles and music over the years.

www.ingramcontent.com/pod-product-compliance
Lightning Source LLC
Chambersburg PA
CBHW072212170526
45158CB00002BA/565